# Milk comes from a COW?

By Dan Yunk

Photographs by John Schlageck
Illustrations by Michele Johnson

BIG
SALE
LOW PRI
MILK

Kailey lives in the city. Her parents are doctors.

She always thought milk came from the grocery store... Until she went to visit Grandma and Grandpa.

MILK
WHEAT

Kailey wanted cereal for breakfast. But when she opened the refrigerator door the milk carton was empty.

Grandma, we need to go to the grocery store. That's where milk comes from.

Grandma decided it was time to show Kailey where milk really came from.

So she loaded Kailey in her car, buckled up, and headed out to a friend's dairy farm.

170
32

On the way there
Grandma told Kailey
milk comes from cows.

All sorts of thoughts
raced through Kailey's
head.

When they got to the dairy farm, Grandma said to her friend farmer George, can you show Kailey where milk comes from?

He showed her his black and white Holstein cows. He told her Holsteins were the most common dairy cows.

Other popular breeds are Guernseys and Jerseys, he said.

George told Kailey all dairy cows must be mothers and give birth to a baby calf before they can produce milk.

Kailey saw the cows eating and drinking.

George said cows eat many times a day. He also told Kailey that cows have a stomach with four chambers!

Every day they eat about 80 pounds of hay, corn and other grasses.

They also drink 30 to 40 gallons of water in a day.

After eating and drinking, they can produce up to 8 gallons of milk daily.

George took Kailey to his milking parlor. She watched the cows come in.

George washed the cows and then attached a milking machine.

The cows' milk is stored in their udders.

The udder is a large bag between a cow's rear legs. A cow has one udder.

The machine gently pulls to get the milk out of the cows' udders.

Milking does NOT hurt the cow.

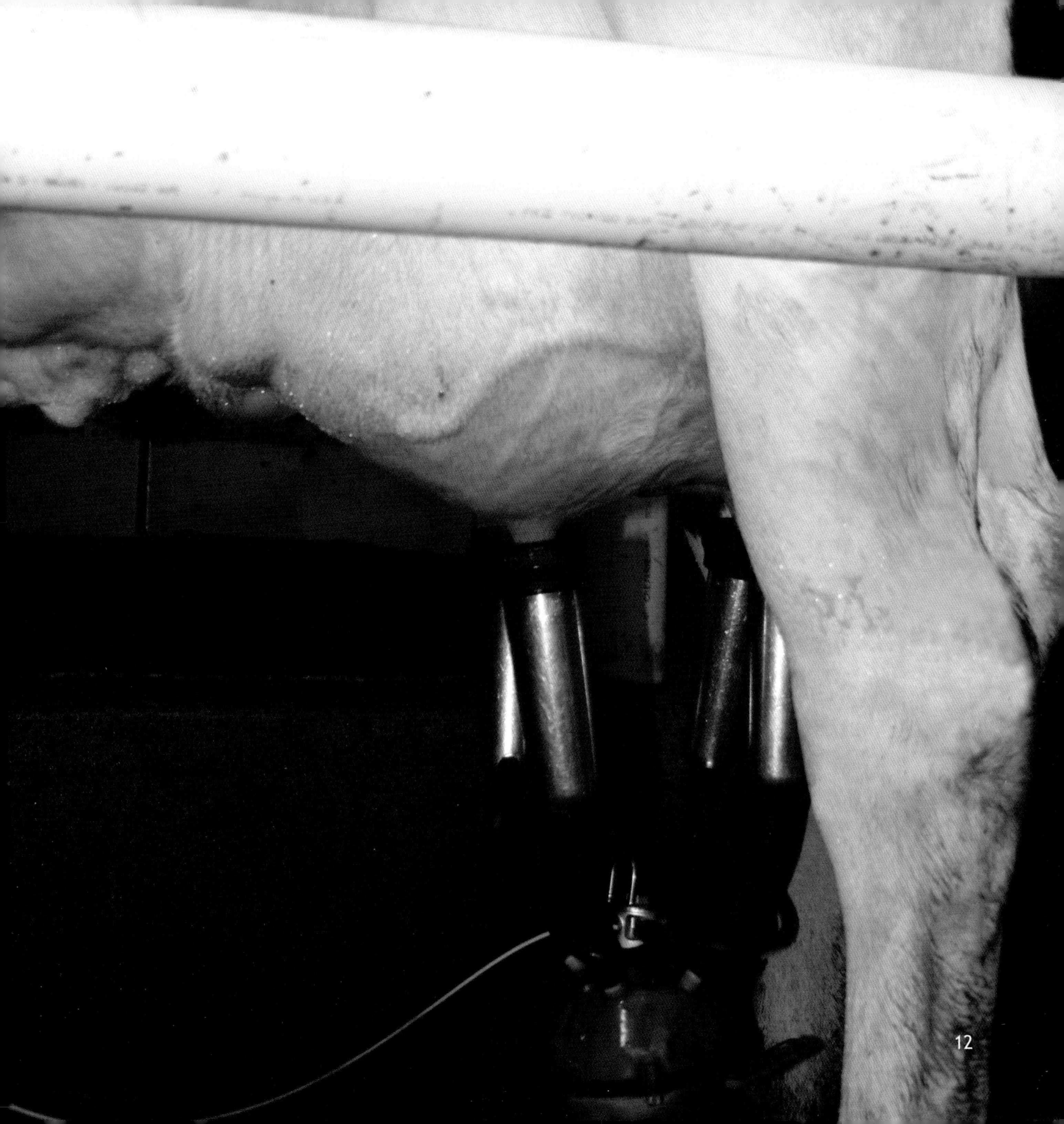

I ♥ MILK

The milking machine pumps the milk from the udder into a large storage tank.

The milk is then cooled and stored and transported by trucks to the processing plant.

The milk is prepared to be put into bottles and cartons.

The bottles and cartons of milk are sent to the grocery stores and finally to our homes, schools, and restaurants.

Does chocolate milk come from brown cows? asked Kailey.

Grandma smiled.

No, all cows give white milk, Grandma said. Chocolate is added to the milk later.

Kailey and Grandma thanked George for showing them that milk comes from a cow.

On her way back to Grandma and Grandpa's house, Kailey was excited.

Grandma, I can't wait to tell my mommy and daddy that milk comes from COWS.